The Farm That Won't Wear Out

by

Cyril G. Hopkins

The Farm That Won't Wear Out
by Cyril G. Hopkins

ISBN: 978-93-61420-27-6

Published by

DOUBLE 9 BOOKS

2/13-B, Ansari Road
Daryaganj, New Delhi – 110002
info@double9books.com
www.double9books.com
Tel. 011-40042856

ABOUT THE AUTHOR

Cyril G. Hopkins, a pioneering figure in the field of agricultural science, cemented his legacy with his seminal work, "The Farm That Won't Wear Out." As an accomplished agronomist, educator, and researcher, Hopkins devoted his life to studying soil fertility and sustainable farming practices. In "The Farm That Won't Wear Out," Hopkins presents a groundbreaking exploration of soil management techniques designed to enhance productivity and preserve the land for future generations. Through meticulous research and practical insights, he offers farmers a comprehensive guide to optimizing crop yields while minimizing environmental degradation.Hopkins' masterpiece delves into the intricacies of soil science, illustrating the importance of maintaining soil health through proper nutrient management, erosion control, and crop rotation. His emphasis on sustainable farming practices resonates with readers, inspiring a deeper appreciation for the intricate relationship between agriculture and the environment."The Farm That Won't Wear Out" stands as a testament to Hopkins' pioneering spirit and enduring commitment to agricultural innovation. His holistic approach to farming serves as a blueprint for sustainable agriculture, guiding farmers towards a more resilient and environmentally conscious future.

CONTENTS

Preface

THE FARM THAT WON'T WEAR OUT

WAS FIRST published serially in THE COUNTRY GENTLEMAN, the privilege having been granted the author of subsequent publication. It is now issued in book form in response to numerous requests coming especially from the Central, Eastern, and Southern States.

CYRIL G. HOPKINS.

CHAMPAIGN, ILL.

> *"Population must increase rapidly, more rapidly than in former times, and ere long the most valuable of all arts will be the art of deriving a comfortable subsistence from the smallest area of soil."—Lincoln.*

> *"It is not the land itself that constitutes the farmer's wealth, but it is in the constituents of the soil, which serve for the nutrition of plants, that this wealth truly consists."—Liebig.*

CHAPTER I
WHAT GOES TO MAKE UP
PERMANENT FERTILITY

IT IS an old saying that "any fool can farm," and this was almost the truth when farming consisted chiefly in reducing the fertility of new, rich land secured at practically no cost from a generous Government. But to restore depleted soils to high productive power in economic systems is no fool's job, for it requires mental as well as muscular energy; and no apologies should be expected from those who necessarily make use of technical terms in the discussion of this technical subject, notwithstanding the common foolish advice that farmers should be given a sort of "parrot" instruction in almost baby language instead of established facts and principles in definite and permanent scientific terms. The farmer should be as familiar with the names of the ten essential elements of plant food as he is with the names of his ten nearest neighbors. Safe and permanent systems of soil improvement and preservation may come with intelligence—never with ignorance—on the part of the landowners.

When the knowledge becomes general that food for plants is just as necessary as food for animals, then American agriculture will mean more than merely working the land for all that's in it. This knowledge is as well established as the fact that the earth is round, although the people are relatively few who understand or make intelligent application of the existing information.

Agricultural plants consist of ten elements, known as the essential elements of plant food; and not a kernel of corn or a grain of wheat, not a leaf of clover or a spear of grass can be produced if the plant fails to secure any one of these ten elements. Some of these are supplied to plants in abundance by natural processes; others are not so provided and must be supplied by the farmer, or his land becomes impoverished and unproductive.

Foods That Plants Live On

Two elements, carbon and oxygen, are contained in normal air in the form of a gas called carbon dioxid, and this compound is taken into the plant through the breathing pores, which are microscopic openings located chiefly on the under side of the leaves. Some plants have more than a hundred thousand breathing pores to the square inch of leaf surface.

When plants or plant products are burned or decomposed the carbon of the combustible material—grass, wood, coal, and so forth—unites with the free oxygen of the atmosphere to re-form the carbon dioxid, which thus returns as a gas to the air. Even the food taken into the animal system, after being digested and carried into the blood, is brought, into contact with the oxygen of the air—which also passes into the blood through the cell walls of the lungs—and a form of combustion takes place, the heat generated serving to warm the body while the carbon dioxid passes back into the lungs and is exhaled into the open air.

By these circulation processes the supply of carbon dioxid in the atmosphere is renewed and maintained without any special effort on the part of man. Hydrogen is one of the elements of which water is composed. Water is taken into the plant through the roots, carried through the stems to the leaves, and there, under the influence of chlorophyll, sunlight and the life principle, the carbon, oxygen and hydrogen are made to unite into some of the most important plant compounds, such as the sugars, which are later transformed into starch and fiber.

Though these three elements constitute the larger part of the mature agricultural plant they are no more necessary for plant growth than the seven which are supplied by the soil. Iron is one of the essential elements of plant food; but the amount required by plants is so small and the amount contained in the soil is so large that soils have never been known to become deficient in iron. Though sulfur is found in plants in very appreciable amounts and is known to be essential to plant growth, it is evident that plants do not need so much sulfur as they often contain, some of it being taken up and merely tolerated, as is the case with all of the sodium and silicon found in plants, neither of these being required for normal growth, although commonly found in plants in very considerable amounts. The supply of sulfur in normal soils is not large; but, with the combustion and decay of organic materials—coal, wood, grass, leaves, and so forth— sulfur passes into the air and is brought back to the soil dissolved in rain or absorbed by direct contact of soil and air. Thus under normal conditions the supply of sulfur naturally provided is ample to meet the needs of the staple farm crops, although there are some plants, such as cabbage, for example, which may possibly be benefited by fertilizing with sulfur.

But there are five other essential elements of plant food, and these require special consideration in connection with permanent soil fertility.

They are potassium, magnesium, calcium, phosphorus and nitrogen. There are also five important points to be kept in mind in relation to each of these elements: (1) the soil's supply, (2) the crop requirements, (3) the loss by leaching, (4) the methods of liberation, and (5) the means of renewal.

The neglect of one or more of these important points in relation to one or more of these five elements has reduced the fertility of most cultivated soils in the United States, has greatly impoverished the older farm lands, and has brought agricultural abandonment to millions of acres in the original thirteen states. On the other hand, intelligent attention to these same factors will bring restoration and high productive power to such lands.

England's Best Lesson in Farming

Where these five elements were supplied regularly to land on the Rothamsted Experiment Station the average yield of wheat for the thirty years, 1852 to 1881, was 35.9 bushels an acre, while 13.6 was the average yield of similar unfertilized land; and during the next thirty years—1882 to 1911—the corresponding average yields were 38 bushels an acre on the fertilized land, and 11.7 bushels where no plant food was applied. These statements are not mere opinions, but determined facts whose accuracy stands unquestioned.

On another field at Rothamsted, England, the average yield of barley for the same sixty years was 43 bushels an acre where nitrogen, phosphorus and calcium were regularly applied, 42.6 where all five elements—including potassium and magnesium—were added, but only 14.3 on unfertilized land.

On still another Rothamsted experiment field, where a four-year crop rotation of turnips, barley, clover (or beans) and wheat has been practiced since 1848, the yield of turnips in 1908 was 717 pounds an acre on unfertilized land and 35,168 pounds where the five important elements of plant food had been regularly applied once every four years—for the turnips only—since 1848. In 1909 the barley yielded 33.4 bushels an acre on the fertilized land, but only 10 bushels where no plant food was applied. The yield of clover in 1910 was 8590 pounds an acre on the land fertilized for turnips, but only 1949 on the unfertilized land. The wheat following the clover with no other fertilizer produced 24.5 bushels an acre in 1911, but 38 bushels where plant food is always applied for turnips grown three years before.

These are the established facts from the longest accurate record, and thus the most trustworthy data the world affords; and when one hears promulgated the very pleasing doctrine that the rotation of crops will maintain the fertility of the soil it is time to remember that "to err is human."

Fertility in Normal Soils

Of the four important mineral elements, potassium is by far the most abundant in common soils. Thus, as an average of ten residual soils from ten different geological formations in the eastern part of United States, two million pounds of subsurface soil were found to contain:

Potassium 37,860 pounds

Magnesium 14,080 pounds

Calcium 7,810 pounds

Phosphorus 1,100 pounds

Even the depleted, and to some extent abandoned, gently undulating upland "Leonardtown loam," which was farmed for generations and which, according to the surveys of the Federal Bureau of Soils, covers 41 per cent of St. Mary's County, Maryland, and more than 45,000 acres of Prince George's County—still contains in two million pounds of surface soil—corresponding to the plowed soil of an acre about 6-2/3 inches deep:

Potassium 18,500 pounds

Magnesium 3,480 pounds

Calcium 1,000 pounds

Phosphorus 160 pounds

The brown silt loam prairie soil of the early Wisconsin glaciation is the most common type of the greatest soil area in the Illinois Corn Belt. Two million pounds of this surface soil contain as an average:

Potassium 36,250 pounds

Magnesium 8,790 pounds

Calcium 11,450 pounds

Phosphorus 1,190 pounds

The older gray silt loam prairie, the most extensive soil of Southern Illinois, contains in two million pounds of soil:

Potassium 24,940 pounds

Magnesium 4,690 pounds

Calcium 3,420 pounds

Phosphorus 840 pounds

These data represent averages involving hundreds of soil analyses, and they emphasize the fact that normal soils are rich in potassium and poor in phosphorus. This is to be expected, for most soils are made from the earth's crust, and normal soils should bear some relation in composition to the average of the earth's crust, which contains in two million pounds 49,200 pounds of potassium and 2,200 pounds of phosphorus, as shown by the weighted averages of analyses involving about two thousand samples of representative rocks, reported by the United States Geological Survey.

Measuring Fertility Losses

The plant food required for one acre of wheat yielding 50 bushels, one acre each of corn and oats yielding 100 bushels, and one acre of clover yielding four tons, includes for the total crops:

Potassium 320 pounds

Magnesium 68 pounds

Calcium 168 pounds

Phosphorus 77 pounds

If only the grain, including a yield of 4 bushels an acre of clover seed, is considered, the straw, stalks and hay being returned to the soil—either directly or in farm fertilizer—then the loss per acre from four years of cropping as above would be as follows:

Potassium 51 pounds

Magnesium 16 pounds

Calcium 5 pounds

Phosphorus 42 pounds

The average annual loss by leaching from good soils in humid sections is known by the results of many analyses to be about as follows per acre:

Potassium 10 pounds

Calcium 300 pounds

Phosphorus 2 pounds

The average annual loss of magnesium in drainage water from good soils is probably 30 pounds or more an acre, but the data thus far secured are inconclusive with respect to that element.

A careful consideration of the trustworthy data clearly reveals the fact that potassium is very abundant in normal soils, while phosphorus is relatively very deficient; and, all things considered, calcium—and probably magnesium—is of much greater significance than potassium, from the standpoint of the maintenance of usable plant food in the soil. It should be noted, too, that certain crops which are exceedingly important for economic systems of permanent agriculture require very large amounts of calcium as plant food. Thus a four-ton crop of clover hay takes about 120 pounds of calcium from the soil, or the same amount as of potassium; while such a crop of alfalfa requires about 145 pounds of calcium, but only 96 pounds of potassium. When it is known that the abandoned "Leonardtown loam" still contains in two million pounds of surface soil 18,500 pounds of potassium and only 1000 pounds of total calcium, the significance of these chemical and mathematical data must be apparent.

The Liberation of Fertility

Probably there has never been a greater waste of time and effort in the name of science than in the endeavor to determine the "available" plant food in soils. The almost universal assumption has been that the plant food in the soil exists in two distinct conditions, "available" and "unavailable," and that the determination of the "available" plant food would reveal both the crop-producing power of the soil and the fundamental fertilizer requirements for the improvement of the soil for crop production.

After ascertaining the total stock of plant food in the plowed soil, the next important question is not how much is "available," but rather how much can be made available during the crop season, year after year. In other words we must make plant food available by practical methods of liberation, by converting it from insoluble compounds into soluble and usable forms; for plant food must be in solution before the plant can take it from the soil. For the present, space is taken only to emphasize the value of decaying organic manures in the important matter of making plant food available; and attention is also called to the fact that the decomposition of the organic matter of the soil—including both fresh materials and old humus—is hastened by tillage and by underdrainage, which permit the oxygen of the air to enter the soil more freely, oxygen being a most active agent in nitrification and other decomposition processes of organic matter, as well as in the more common combustion of wood, coal, and so forth.

The Renewal of Fertility

In rational systems of general farming the supply of any element which is normally very abundant may be renewed from the subsoil by even the very slight erosion which occurs on all ordinary lands in humid sections. This statement applies to iron and potassium, and often to magnesium.

If two million pounds of normal surface soil contain 30,000 pounds of potassium, one inch an acre would contain 4500 pounds of that element; and if a third of this—1500 pounds—were removed by cropping and leaching before its removal by surface washing, then two-thirds of a century could be allowed for the erosion of one inch of soil, with crop yields of 50 bushels of wheat, 100 bushels of corn and oats, and 4 bushels of clover seed to the acre, provided the stalks, straw and clover hay were returned to the land, either directly or in farm manure. This amount of surface washing is likely to occur on land sufficiently undulating for good surface drainage, provided the land is plowed and cultivated as frequently as would be required for a four-year rotation as suggested above. Where hay, straw, potatoes, root crops or common market garden crops are sold, very much larger amounts of potassium leave the farm than in grain farming or live-stock farming, and in such cases potassium must ultimately be purchased and returned to the soil, either in commercial form or in animal manures from the cities.

Thirty Bushels for Potassium

There are some soils, however, which are not normal—soils whose composition bears no sort of relation to the average of the earth's crust; such, for example, as peaty swamp soil or bog lands, which consist largely of partly decayed moss and swamp grasses. These soils are exceedingly poor in potassium, and they are markedly and very profitably improved by potassium fertilizers, such as potassium sulphate and potassium chloride— commonly but erroneously called "muriate" of potash.

Thus, as an average of triplicate tests each year, the addition of potassium to such land on the University of Illinois experiment field near Manito, Mason county, increased the yield by 20.7 bushels more corn to the acre in 1902, by 23.5 in 1903, by 29 in 1904 and by 36.8 in 1905; and the proceedings of the midsummer session of the Illinois State Farmers' Institute for 1911 report that the use of $22,500 in potassium salts on the peaty swamp lands in the neighborhood of Tampico, Whiteside county, increased the value of the corn crop in 1910 by $210,000, the average increase for potassium being about 30 bushels of corn to the acre.

Some sand soils, particularly residual sands, which often consist largely of quartz-silicon dioxid—are very deficient in potassium; consequently the experiments or demonstrations conducted by the potash syndicate at Southern Pines, North Carolina, show very marked increases from the use of potassium salts on such soil, although the result ought not to be used to encourage the use of such fertilizers on normal soils, which are exceedingly rich in potassium.

Even in soils abundantly supplied with potassium temporary use may well be made of soluble potassium salts when no adequate supply of decaying organic matter can be provided. For this purpose, kainit—which contains potassium and also magnesium and sodium in chlorides and sulfates—is preferred to the more concentrated and more expensive potassium salts. About 600 pounds an acre every four years is a good application. The kainit will not only furnish soluble potassium and magnesium but will also help to dissolve and thus make available other mineral plant food naturally present or supplied, such as natural phosphates. When the supply of organic matter produced in crops and returned either in farm manure or in crop residues becomes sufficiently abundant, then the addition of kainit may be discontinued on normal soil.

Thus, as an average of 112 separate tests covering four different years, on the Southern Illinois experiment field on worn, thin land, at Fairfield, the use of 600 pounds an acre of kainit once in four years increased the yield of corn by 10.7 bushels where no organic manure was used, and by only 1.7 bushels when applied with eight tons of farm manure.

Liming the Soil

In the form of ashes, marl or chalk, lime has been used as a fertilizer for thousands of years. It serves two very important purposes: to correct the acidity of sour soils and to supply calcium and sometimes magnesium as plant food. Burned lime has also been much used, but in more recent years the development of machinery for crushing and pulverizing rock— especially in cement manufacture—has made possible the production of pulverized natural limestone, and at much less expense than for caustic lime made by burning and slaking. Where ground limestone can be easily procured it takes the place of burned lime, and it produces better results at less expense, even though 1-3/4 tons of ground limestone are required to equal 1 ton of quicklime in calcium content and in power to correct acidity.

Furthermore, ground limestone can be applied in any amount with no injurious results, while caustic lime destroys the organic matter or humus of the soil, dissipates soil nitrogen, is disagreeable to handle, and may injure the crop unless applied in limited amounts or several months before the crop is to be planted.

The most valuable and trustworthy investigation on record in regard to the comparative value of burned lime and ground limestone has been conducted by the Pennsylvania Experiment Station. A four-year rotation of crops was practiced, including corn, oats, wheat and hay (clover and timothy) on four different fields, each crop being represented every year. After twenty years the results for the four acres showed that the land treated with ground limestone had produced 99 bushels more corn, 116 bushels more oats, 13 bushels more wheat and 5.6 tons more hay than the land treated with about an equivalent amount of burned lime. At the end of sixteen years the analysis of the soil showed that the burned lime had destroyed 4.7 tons of humus and had dissipated 375 pounds of nitrogen to the acre, as compared with the ground limestone, this loss being equivalent to 37-1/2 tons of farm manure.

Other trustworthy experiments by the Maryland and Ohio Experiment Stations confirm the Pennsylvania results in showing better crop yields when unburned lime carbonate was used; and more extensive experiments by the Tennessee Experiment Station also agree with the Pennsylvania data in regard to the destruction of organic matter and loss of soil nitrogen from the use of burned lime. If dolomitic limestone is used, magnesium as well as calcium is thus added to the soil.

Limestone need not be very finely pulverized. If ground so that it will pass through a ten-mesh sieve it is amply fine, assuming that the entire product is used, including the finer dust produced in grinding, and it is very possible that final investigations will show that the entire product from a quarter-inch screen is even more economical and profitable in permanent systems.

Limestone is quite easily soluble in soil water carrying carbonic acid. It is thus readily available; in fact, it is too available to be durable if very finely ground; and in humid sections the loss by leaching far exceeds that removed by cropping. In practical economic systems of farming about two tons an acre of ground limestone should be applied every four years, or corresponding amounts for other rotation periods.

The essential facts relating to potassium, magnesium and calcium and to the use and value of different forms of lime have been stated above, and they may be accepted with confidence for use in economic systems of farming on normal soils.

CHAPTER II
THE NITROGEN PROBLEM AND
ITS ECONOMICAL SOLUTION

IN THE previous chapter emphasis has been laid upon the fact that plants as well as animals must have food, and that the neglect or ignorance of this factor in American agriculture has led to soil depletion and land ruin on vast areas, especially in the older states.

It has been shown that of the ten essential elements of plant food, five are provided by natural processes without the intervention of man; that, of the remaining five, potassium is the most abundant in normal soil, but requires liberation by good systems of farming; that ground natural limestone is the ideal material with which to supply calcium and to prevent or correct soil acidity; and that if dolomitic limestone be used magnesium is also supplied in suitable form for plant food, Thus only nitrogen and phosphorus remain for consideration.

Keeping in mind that systems of permanent profitable agriculture in America must be founded upon an intelligent understanding of the foundation principles involved, let us pray for strength to acknowledge the truth and cease trying to deceive ourselves. The truth is that by soil enrichment alone the average crop yields of the United States could be doubled, with the same seed and seasons and with but little more work than is now devoted to the fields; and we should cease trying to deceive ourselves in the hope or belief that the fertility of our soil will be maintained if we continue year after year to take crops from the land and fail to make adequate return.

Nitrogen is both the most abundant agriculturally and the most expensive commercially of all the elements of plant food; and yet there is a method by which it can be secured not only without money but with profit in the process. The percentage of nitrogen in normal soils decreases with depth, so that subsoils are almost devoid of nitrogen. This would be more generally understood if it were known that the supply of soil nitrogen in humid countries is contained only in the organic matter.

This organic or vegetable matter consists of the partly decomposed residues of plants, including the roots and fallen leaves which may accumulate naturally, and the green manure crops, crop residues and farm manure which may be supplied in farm practice. Thus the nitrogen of a soil is measured approximately by its content of organic matter; and, vice versa, the percentage of nitrogen is an approximate measure of the organic matter, because nitrogen is a regular constituent of the organic matter normally contained in soils. Consequently if the organic matter of a soil is reduced the supply of nitrogen is also reduced.

In the most depleted soils nitrogen is usually the most deficient element, although it may not be the only deficiency. Thus in the depleted "Leonardtown loam," which occupies such extensive areas of land in Southern Maryland, near the District of Columbia, and which has been to a large extent agriculturally abandoned after one or two centuries of farming, only 900 pounds of nitrogen are found in the plowed soil of an acre—that is, in 2,000,000 pounds of surface soil, corresponding to about 6-2/3 inches an acre. This total amount if made available would be sufficient for only six such crops of corn as are actually produced on our best land in good seasons, and yet it is four times as much as is contained in an equal weight of the subsoil.

The average prairie land of the Corn Belt contains only 5000 pounds of nitrogen in the plowed soil of an acre 6-2/3 inches deep, whereas a 100-bushel crop of corn removes 150 pounds of nitrogen from the soil. A simple computation shows the supply in the plowed soil to be sufficient for only 33 such crops. Even the 100-bushel crop of corn per acre is known to have been produced in many places on exceptionally rich land, and yet the ten-year average yield in the United States is only 25 bushels to the acre.

200 Per Cent for Nitrogen

On Broadbalk Field at Rothamsted, England, wheat has been grown on the same land every year for about two-thirds of a century. As an average of the sixty years, 1852 to 1911 the yield was 12.6 bushels an acre on unfertilized land, 14.6 where mineral plant food was annually applied, 20.3 where nitrogen salts alone were used, and 37 where both nitrogen and mineral plant food were applied.

During the thirty years, 1882 to 1911 the average yields were 11.7 bushels an acre on the unfertilized land, 14 with minerals, 18.7 where only nitrogen salts were used, and 38 where both nitrogen and minerals were regularly supplied.

These absolute data from the oldest agricultural experiment station in the world should help us to understand why the ten-year average yield of wheat is 33 bushels an acre for all of Great Britain, 37-1/2 for England alone, and only 14 for the United States.

The application of nitrogen increased the yield of wheat by 24 bushels an acre—from 14 to 38 bushels—as an average of the last thirty years, following an average increase of 26.3 for the nitrogen applied during the previous thirty years. It is true that the cost of the fertilizers used exceeded the value of the increase in yield; but let us bear in mind that this truth does not destroy the other truth.

Prove all things, and hold fast that which is good. It is a good fact that 1218 bushels of wheat were produced by the application of nitrogen to an acre of land during a period of sixty years, over and above the produce of another acre which differed only by not receiving nitrogen; whereas the total produce from an acre of unfertilized land was only 756 bushels during the same sixty years. It is a good fact that the increase alone from the nitrogen applied is more than twice the total yield of the unfertilized land during the last thirty years, and he does well who holds fast this fact.

It is also a good fact that as an average of sixty years the yield of barley was increased by 21.6 bushels an acre by nitrogen; that nitrogen increased the yield of hay on permanent meadow land at Rothamsted by 1-1/2 tons an acre as a fifty-year average; and that nitrogen increased the average yield of potatoes by 88 bushels as an average of twenty-six years; while the average of the unfertilized land was only 51 bushels an acre, these increases in barley, bay, and potatoes being obtained over and above the yields where minerals alone were used.

Where Is Nitrogen?

If nitrogen has such enormous power to increase the yield of our great staple farm crops then we may well inquire, Where is nitrogen, and how can it be secured economically and utilized profitably in practical agriculture?

The weight of the atmosphere is 15 pounds to the square inch. This means that a column of air 1 inch square taken to the full height of the terrestrial atmosphere weighs 15 pounds. More than three fourths of the air is nitrogen. Since there are 43,560 square feet in one acre, it follows that the nitrogen in the air above each acre of the earth's surface amounts to 70,000,000 pounds, or nearly 500,000 times the 150 pounds of nitrogen required for a hundred-bushel crop of corn. The leaves of the corn plant are blown about by the wind carrying 75-1/2 per cent of nitrogen, but cannot utilize an ounce of this supply.

Many people know that clover and other legumes have power, through the bacteria which inhabit their root tubercles, to feed upon the inexhaustible supply of atmospheric nitrogen which freely enters the pores of the soil; but who knows how much nitrogen is taken from the air by a given crop of clover? Not one in a thousand can answer this question; and meanwhile our continued agricultural and national prosperity depends in large part upon the possibility of wide dissemination and practical application of a quantitative knowledge of the nitrogen problem.

As a rule the so-called "practical" farmer is a theorist. He first believes that the virgin soil is inexhaustible, even though cropped continuously. Later he clings to the popular theory that the rotation of crops will maintain the productive capacity of the land; and it is safe to say that a large majority of the farmers of the United States gladly hold to the erroneous theory that clover grown once every three to five years will increase and permanently maintain the fertility of the soil.

The fact that clover was grown for generations on the lands of the older Eastern states until the clover crop itself finally failed on millions of acres now agriculturally abandoned is overlooked or forgotten by present-day farmers, especially by the descendants of those who have gone West and settled on new, rich lands.

Six Facts and a Question

The following six facts will furnish a comprehensive basis for the solution of the nitrogen problem in practical general agriculture:

(1) To produce 100 pounds of grain requires about 3 pounds of nitrogen, of which 2 pounds are deposited in the grain itself and 1 pound in the straw or stalks.

(2) In live-stock farming one-fourth of the nitrogen in the food consumed is retained in the animal products—meat, milk, wool, and so on—and three-fourths may be returned to the land in the excrements if saved without loss.

(3) When grown on soils of normal productive capacity legumes secure about two-thirds of their total nitrogen from the air and one-third from the soil.

(4) Clover and other biennial or perennial legumes have about two-thirds of their total nitrogen in the tops and one-third in the roots, while the roots of cowpeas and other annual legumes contain only about one-tenth of their total nitrogen.

(5) Hay made from our common legumes contains about 40 pounds of nitrogen per ton.

(6) Average farm manure contains 16 pounds of nitrogen per ton.

Question: How many tons of average farm manure must be applied to a 40-acre field in order to provide as much nitrogen as would be added to the soil by plowing under 2-1/2 tons of clover per acre? Answer: 400 tons.

Either method will furnish about as much nitrogen as would be taken from the soil by a 50-bushel crop of wheat, a 75-bushel crop of corn or a 100-bushel crop of oats per acre. The decision by the individual between live-stock farming and grain farming should be based upon preference and profit rather than upon the erroneous teaching that farm manure is either essential or sufficient for the maintenance of soil fertility in this country.

Bread is the staff of life, and many must sell grain. I do not advise all grain farmers to become live-stock farmers; but I do advise both grain farmers and live-stock farmers to enrich their soils by practical, profitable and permanent methods. Both classes of farmers may secure new nitrogen—that is, they can positively increase their nitrogen supply by sufficient use of legume crops.

How to Supply Nitrogen

The cotton-grower who sells cotton lint at 10 cents a pound and the market gardener who sells from $100 to $300 worth of fruits and vegetables from one acre may well make liberal use of commercial nitrogen at 15 or 20 cents a pound; but if after deducting the cost of harvesting, threshing, storing and marketing the average farmer receives only 1 cent a pound for his grain and if 40 per cent of the commercial nitrogen applied is lost by leaching, then the total crop of grain would bring only enough money to pay for the nitrogen required to produce it, at 20 cents a pound. We may sometimes advise the American grain-grower to buy water with which to irrigate his crop, but not to buy nitrogen with which to fertilize it.

If the grain farmer grows 40 bushels of wheat to the acre, clover having been seeded on the same land in order to plow under the equivalent of 1-1/2 tons of hay as green manure the following spring, and follows this by a 60-bushel crop of corn and a 50-bushel crop of oats, and this the fourth year by two crops of clover aggregating 3 tons an acre, including 2 bushels of seed, he can thus secure from the air about 180 pounds of nitrogen in the 4-1/2 tons of clover. Moreover, if the first cutting of clover the fourth year is left on the land and the threshed clover straw from the seed crop and likewise all straw and stalks are returned to the soil, only 154 pounds of nitrogen an acre would leave the farm if the total grain and clover seed were sold. With 80 cents a bushel for wheat, 50 cents for corn, 40 cents for oats and $8 for clover seed, the total returns from the four acres would amount to $98.

On the other hand the live-stock farmer may grow two 60-bushel crops of corn, followed by 50 bushels of oats and then 3 tons of clover hay containing 120 pounds of new nitrogen. The four crops would contain 350 pounds of nitrogen; and if the grain and hay and half the corn-stalks are used for feed, with the straw and the remainder of the stalks for bedding, it is likewise possible to replace the 230 pounds of nitrogen required for the grain crops, provided not more than one-seventh of the manure is lost before being returned to the land. The important weakness on the common live-stock farm lies in the enormous waste of manure.

If 10 pounds of feed produce 1 pound increase in the live-weight of the animals fed, and if they bring 6 cents a pound on the hoof, the gross returns aggregate $107.50 from the four acres, barring losses from accidents, animal diseases, and so on.

Thus, with a few established facts in mind, one can easily determine how to maintain or even to increase the supply of nitrogen in the soil, and without the purchase of nitrogen in any form; and it is just as possible and just as necessary thus to provide the nitrogen needed in grain farming as in livestock farming. When we consider that animals destroy two-thirds of the organic matter in the food consumed we find that as between the two systems above described the organic matter or humus of the soil will be better maintained in the grain system outlined.

Live-Stock or Grain Farming

For those who believe that live-stock farming must be adopted for the maintenance of fertility on all farms, attention should be called to the fact that there are 900,000,000 acres of farm-land in the United States and only 90,000,000 head of live-stock equivalent to cows, including all farm animals. Will the manure from one cow serve to enrich 10 acres of land?

It should also be known that a hundred bushels of grain will support five times as many people as could live for the same length of time on the meat and milk that could be made by feeding the grain to domestic animals. It is because of this fact that the consumer may sometimes boycott meat or other animal products, while he never boycotts bread; but let us hope that permanent systems will become generally adopted in America, for the production of both grain and live stock, so that high standards of living may be maintained for all classes of people in this country.

The oldest direct comparison between these two systems of farming, so far as the writer has learned, is on the experiment fields of the University of Illinois, where as an average of six years the yield of corn has been 87 bushels an acre in grain farming and 90 bushels in live-stock farming, the same crop rotation being practiced. Where wheat was introduced the average yield for six years was 43.1 bushels in grain farming and 43.5 in live-stock farming.

No nitrogen was purchased in any form in either of these systems; but clover is grown in the rotation to secure nitrogen from the air and then the crop residues or farm manure is returned to the soil to provide sufficient nitrogen for the grain crops. In all cases phosphorus was used for these yields.

Even more encouraging than these six-year average results from Illinois are the results of sixty years from Agdell Field at Rothamsted.

Where mineral plant food was regularly applied, and where all the manure produced by feeding the turnips was returned to the soil, in a four-year rotation of turnips, barley, clover (or beans) and wheat, with no other provision made for supplying nitrogen, the yields per acre were as follows:

Turnips, 24,724 lbs. in 1848, and 26,410 in 1908.

Barley, 42.8 bushels in 1849 and 22.1 in 1909

Clover, 5586 pounds in 1850 and 7190 in 1910.

Wheat, 32 bushels in 1851 and 37.8 in 1911.

Here we have data which span a period of sixty years and which show that where mineral plant food has been provided the clover in rotation and the manure produced by the feeding of only one of the four crops have maintained the yield of all crops except the barley-the third crop after clover-and without the application of nitrogen in any other form. If the clover and straw had been returned to the land either directly or in farm manure the additional nitrogen thus provided would have been sufficient both to maintain the yield of barley and to prevent the moderate decrease which has occurred in the nitrogen content of the soil.

CHAPTER III
PHOSPHORUS: THE MASTER KEY
TO PERMANENT AGRICULTURE

THE greatest economic loss that America has ever sustained has been the loss of energy and profit in farming with an inadequate supply of phosphorus. Phosphorus is a Greek word which signifies "light-bringer"; but it is a light which few Americans have yet seen, else we should not permit the annual exportation of more than a million tons of our best phosphate rock, for which we receive at the mines the paltry sum of five million dollars, carrying away from the United States an amount of the one element of plant food we shall always need to buy, which if retained in this country and applied to our own soils would be worth not five million but a thousand million dollars for the production of food for the oncoming generations of Americans.

For five million dollars we export to Europe each year enough phosphorus for 1,400,000,000 bushels of wheat, or twice the average crop of the entire United States. Meanwhile our ten-year-average yield of wheat is 14 bushels an acre, while Germany's yield has gone up to 29, Great Britain's to 33, England's to 37-1/2 and Denmark's to more than 40 as the average for a decade.

Potato Yield Twice Doubled

There is only one place in the world where we can go for the results of soil improvement for more than a quarter of a century in connection with the growing of potatoes. Of course this place is Rothamsted, England, where as an average for twenty-six years the yield of potatoes was 51 bushels an acre on unfertilized land and exactly 102 bushels where only a phosphate fertilizer was applied. Where the same amount of phosphorus—29 pounds of the element per acre per annum—was used in connection with other minerals—300 pounds of potassium sulfate and 100 pounds each of the sulfates of magnesium and sodium—the average yield of potatoes was 109 bushels. Where 86 pounds of nitrogen was applied in sodium nitrate the average yield was 79 bushels; but where the nitrogen, phosphorus and other minerals were all applied the average yield for the twenty-six years was 203 bushels.

At 50 cents a bushel for potatoes, the investment in phosphorus alone paid 600 per cent net profit; and even the complete fertilizer, including 392 pounds of acid phosphate, 550 pounds of sodium nitrate and 500 pounds of alkali salts, aggregating 1442 pounds, and costing at moderate prices $24.28 an acre per annum, paid back $76 a year as a twenty-six year average, thus making 300 per cent even on an investment of nearly $25 an acre a year.

Phosphorus Helps Good Farming

There is also but one place in the world where we can learn the results secured from the application of phosphorus for a period of thirty-six years in a good system of farming; and again this place is Rothamsted.

In 1848 Sir John Lawes and Sir Henry Gilbert began investigations on Agdell Field. The Norfolk rotation, already known at that time as one of the best rotation systems, was turnips, barley, clover, and wheat; and in these practical field experiments the turnips were fed on the land and the animal fertilizer thus produced was returned to the soil, which was well supplied with limestone.

During the next thirty-six years $29.52 worth of phosphorus per acre was applied to one part of the field; and in comparison with another part of the same field cropped and managed similarly, except that no phosphorus was applied, the $29.52 worth of phosphorus produced $98.02 increase in the value of the turnips, $37.45 in barley, $48.93 in clover (and beans) and $45.99 in wheat.

The total value of the crops grown on the land not receiving phosphorus during the thirty-six years was $432.43 an acre, while on the phosphated land the crop values amounted to $662.82, an increase of $230.39 from an investment of $29.52, the turnips being figured at $1.40 a ton, barley at 50 cents a bushel, clover hay at $6 a ton, beans at $1.25 a bushel, wheat at 70 cents a bushel, and phosphorus at 12 cents a pound. As a general average at these conservative prices, the investment of $3.28 an acre every four years paid back $25.60 in the four crops.

In most states the legal rate of interest is 6 per cent but here is an investment that paid the principal and 680 per cent interest every four years. And these investigations show that the phosphorus was used with profit for the production of markedly different crops, including potatoes and turnips, barley and wheat, clover and beans.

But the soil at Rothamsted is no poorer in phosphorus than is the average soil of the United States; and these results are given here not only because they are the oldest and most trustworthy the world affords, but because they are strictly applicable to the production of common crops on vast areas of agricultural land in our own country.

The Form of Phosphorus to Use

The unfertilized soil at the Rothamsted station contains, in 2,000,000 pounds—corresponding to about 6-2/3 inches to the acre—1000 pounds of phosphorus and 35,000 of potassium, while an acre of plowed soil of the same weight at State College, Pennsylvania, contains 1100 pounds of phosphorus and 50,700 of potassium.

In a word, normal soils are deficient in phosphorus, and the application of phosphorus in good systems of farming produces marked and profitable increases in crop yields. But what form of phosphorus shall we apply? This is a very important question in agricultural economics, for we have many different kinds of fertilizing materials that contain phosphorus, and one may cost ten times as much as another as a source of phosphorus. Thus 250 pounds of phosphorus in a ton of finely ground natural rock phosphate can be purchased at the mines in Tennessee and delivered at the farmer's railway station in the heart of the Corn Belt for $8. Or the ton of raw phosphate may be mixed with a ton of sulfuric acid in the fertilizer factory, and the two tons of acid phosphate may be sold to the same farmer for $32. Or the fertilizer manufacturer may mix the two tons of acid phosphate with two tons of "filler," containing a little nitrogen and potassium, and then sell the same farmer the four tons of so-called "complete" fertilizer for $80; and the farmer gets no more phosphorus in the four tons of "complete" fertilizer for $80 than in the one ton of natural phosphate for $8.

The Pennsylvania State College conducted an experiment for twelve years—1884 to 1895—in which $1.05 an acre was invested in ground raw rock phosphate with a rotation of corn, oats, wheat and hay (clover and timothy), and the value of the increase produced by the phosphorus amounted to $5.85 as an average for the twelve years, and to $8.41 as an average for the last four years. Thus the profit was from about 560 to 800 per cent on the investment, counting corn at 35 cents a bushel, oats at 30 cents, wheat at 70 cents, and hay at $6 a ton. These figures represent the increase produced by phosphorus over and above the value of the crops grown without phosphorus fertilizer. In this case no farm manure was used on either part of the field; but commercial nitrogen and potassium were applied alike on both parts, and clover was grown in the rotation.

Acid phosphate was also used in direct comparison; and, in answer to the question whether the general farmer should apply liberal amounts of finely ground natural rock phosphate, or whether he should pay four times as much for phosphorus after the fertilizer manufacturer has mixed one part of the raw rock with one of sulfuric acid and thus produced two parts of acid phosphate, these Pennsylvania experiments tell us that the yearly average for the twelve years gave a gain per year of $2.45 from the raw phosphate and 48 cents from the acid phosphate, at the prices used by the Pennsylvania Experiment Station. But we must not draw general conclusions from this one experiment, even though it covers twelve years.

In 1895 the Maryland Experiment Station began field experiments with different forms of phosphorus; and, as an average of six tests every year for twelve years, $1.965 invested in ground raw rock phosphate produced increases in corn, wheat and hay that were worth $22.11, at 35 cents a bushel for corn, 70 cents for wheat, $6 a ton for hay, and 3 cents a pound for phosphorus in the ground natural phosphate. How would you like 1000 per cent profit as the result of mixing brain with brawn, in connection with the improvement of your own business, thus keeping the investment under your own control?

Mind you, this does not prove that farming is profitable, but only that the intelligent use of phosphorus in farming is profitable. In other words the admixture—brains—is profitable.

In commenting upon his investigations the director of the Maryland Agricultural Experiment Station states that the raw phosphate produced a higher total average yield than acid phosphate, and at less than half the cost.

The Rhode Island Experiment Station began a series of experiments with different forms of phosphorus in 1894. If we add together all the hay and grain crops grown during the decade following the first year of these experiments, we find that the increases per acre were 14,580 pounds for raw phosphate and 14,550 pounds for acid phosphate, on unlimed land; while lime and raw phosphate produced 27,030 pounds, and lime and acid phosphate 29,690 pounds, of increase; and the acid phosphate cost three times as much as the raw phosphate.

In commenting upon these investigations the director of the Rhode Island Experiment Station states that the raw phosphate gave very good results with such farm crops as oats, peas, crimson clover, millet, soy beans, and so forth, but very poor results with such garden crops as turnips, rutabagas, cabbage, beets, lettuce, squash, and so forth, and its use for these garden crops is not advised.

In 1890 the Massachusetts Experiment Station began investigations with different phosphates applied in equal money value, and in his report for 1900 the director states that the raw rock phosphate ranks above the acid phosphate both as an average for the entire period and as an average between 1895 and 1900, during which time the land to which no phosphorus was applied produced only 55 per cent as much as where raw phosphate was used—a result worth every farmer's consideration.

More Bushels and Tons

The Ohio Agricultural Experiment Station has reported investigations covering sixteen years in which raw phosphate was compared with acid phosphate costing twice as much per acre. As an average of all results secured, 320 pounds of raw phosphate applied with manure on clover sod produced 8.4 bushels more corn, 4.7 bushels more wheat, and 0.49 ton more hay per acre than where manure alone was used, and 320 pounds of acid phosphate, costing twice as much money but containing only half as much phosphorus, applied with the same amount of manure, produced 7.5 bushels more corn, 5.1 bushels more wheat, and 0.46 ton more hay than where the manure alone was used.

Now I have presented the averages or summaries of all investigations that have been reported covering ten years or more where equal money values of raw phosphate and acid phosphate have been used, or where any apparent provision was made to supply some organic manure, whether as farm manure, green manure or merely as clover grown in the rotation; and I invite the reader to mix his own brains with these data and not to expect me to state whether he should use the relatively cheap ground natural phosphate rock or the more costly manufactured acidulated phosphate in the improvement of his own soil in systems of permanent profitable agriculture.

Making Phosphate Available

If the natural rock is used it should be ground so that at least 90 per cent will pass through a sieve with 10,000 meshes to the square inch, and of course its content of phosphorus (from 12 to 15 per cent) or of so-called "phosphoric acid" (from 27 to 34 per cent) should also be guaranteed. Moreover it should be used liberally and in connection with plenty of decaying organic matter. People sometimes ask, "How much of the phosphorus in raw phosphate is available?" The best answer to this question is, "None of it; and, if you are not going to make it available, don't use it."

On my own farm I use about one ton per acre of raw phosphate once every six years, thus adding at least 250 pounds of phosphorus at a cost of less than $8; whereas 200 pounds of the common "complete" fertilizer per acre yearly would cost $12 every six years, and would supply only 40 pounds of phosphorus. I do not use "complete" fertilizers, because there is plenty of nitrogen in the air and plenty of potassium in the soil; and because, by growing and plowing under plenty of clover, I not only secure nitrogen from the air and liberate potassium from the soil but also liberate the phosphorus from the raw rock phosphate applied to the soil. In beginning the use of raw phosphate where the supply of organic manures is limited, I apply one ton of phosphate and 600 pounds of kainit in intimate connection, turn them under, preferably with organic matter, then add ground limestone if needed, and thus prepare to grow clover.

By far the most important agencies under the farmer's control for the liberation of plant food are the decomposition products of fermenting or decaying organic matter, such as green manures, crop residues and ordinary farm manures. In the decomposition of these organic materials sour or acid products are formed. Thus vinegar, containing acetic acid, is formed from the fermentation of apple juice, hard cider being an intermediate product. Sweet, chopped, immature field corn becomes sour silage in the silo, lactic, acetic, carbonic and other acids being formed. By a similar process cabbage is turned into sauerkraut. Likewise sweet milk becomes sour, with the formation of lactic acid. Oxalic, citric, tartaric, succinic, malic, gallic and tannic are other well-known organic acids. Some of these are contained in the sap or juice of certain plants, and these or others are formed when crop residues are decomposed in the soil.

In the ultimate decomposition of organic matter the carbon appears in the form of carbon dioxid which when combined with water forms carbonic acid. Though this is a very weak acid, its solvent action is very important.

But, in addition to the various organic acids and carbonic acid, we have also to consider the formation of nitric acid in connection with the decomposition of organic manures. Nitric acid is one of the strongest known, and in solvent power it is excelled by no single acid. The nitrogen contained in crop residues and other organic manures is chiefly in chemical combination with carbon, oxygen and hydrogen, much of it in insoluble protein compounds. Normally this organic nitrogen is transformed in the soil, first into ammonia nitrogen, next into nitrite nitrogen, and lastly into nitrate nitrogen, these three transformations being effected by biochemical action produced by different kinds of living microscopic organisms called bacteria. Though detectable amounts of free nitric acid do not accumulate during this process of nitrification, the soluble nitrate or final product is formed by the action of nitric acid upon a mineral base, such as calcium, magnesium, or potassium, which may have been in the soil in insoluble form, so that the nitrogen must pass through the form of nitric acid in the transformation into nitrates.

While the organic matter applied to the soil contains about twenty times as much carbon as nitrogen, and while corresponding amounts of carbonic acid and important amounts of intermediate organic acids must be formed, it is of much interest to know that even the nitric acid formed in the transformation of organic nitrogen to nitrate nitrogen in sufficient quantity for a given crop is seven times as much acid as would be required to convert raw rock phosphate into soluble phosphate to furnish the phosphorus required for the same crop. A knowledge of this definite quantitative relationship should help us to appreciate the possibilities of decaying organic manures in the important matter of making plant food available, including potassium, calcium and magnesium as well as phosphorus and nitrogen.

The value of rye, rape, buckwheat and other non-legumes when used as green manures is very largely due to the liberation of plant food by their decomposition in contact with the natural phosphates, potash and other minerals contained in the soil. The farmer has no more important business than that of making plant food available, especially by supplying liberal amounts of decaying organic matter.

The following suggestions are offered to the land owner:

To enrich the soil apply liberal amounts of limestone, organic manures and phosphorus.

To enrich the seller apply small amounts of high-priced "complete" commercial fertilizers.

Thus the average of seventy-three "Cooperative Fertilizer Tests on Clay and Loam Soils," extending into thirty-eight different counties in Indiana (Bulletin 155), shows 13 cents as the farmer's profit from each dollar spent for "complete" fertilizers used for corn, oats, wheat, timothy, and potatoes, if valued in the field at 40 cents a bushel for corn, 30 cents for oats, 80 cents for wheat, 50 cents for potatoes, and at $10 a ton for hay, over and above the extra expense for harvesting and marketing the increase, and of course the soil grows poorer, because the crops harvested removed much more plant food than the fertilizers supplied.

CHAPTER IV
PERMANENT SOIL FERTILITY

Its Relation to Profits and Future Values

THOUGH intelligent soil improvement is the most profitable business in which an honest man can engage, ordinary farming is not a highly remunerative occupation, and to a large extent the fortune of the farmer is bound up with the increase or depreciation in the market value of his land. There are at least three important factors of influence which induce people to continue farming:

First, the farmer is his own employer. He controls his own job, is his own boss and has no superior officer to lay him off because of disagreement, dull business or political preferment. Farmers constitute by far the largest class of citizens who own their own business, and are thus "independent."

Second, the farmer is able as a rule to make some sort of a living for his family very largely out of the produce of the farm, so that he gets some return for his labor in terms of food, even when there is no profit in farming as a business; whereas the wage-earner of the city, as soon as his wages stop and his savings and credit are exhausted, must see his family supported by charity or starve. This is not fiction, but fact.

Third, land is usually considered a safe investment, in which one may hold a perfect and undivided title to his property; and people will retain possession of a farm even when it pays a low rate of interest, rather than sell and invest the proceeds in some other enterprise which they cannot control as individuals or which may suddenly depreciate in earning power, fail or be utterly destroyed.

Is Land a Safe Investment?

Though it is true that farm land does not pass out of existence in a day, nevertheless it is by no means a safe investment, as witness the numerous abandoned farms in the older agricultural sections of this new country. It is easily possible for one of means to become land-poor—to have investments in land which will not pay the taxes and upkeep of buildings, fences and so forth. At prevailing prices for farm produce and labor there are vast areas of land in the older states far past the point of possible self-redemption; and, as a matter of business, one might better burn his money and save his energy than to expend all his resources in half-paying for such depleted land, depending upon the immediate income from it to raise a mortgage covering the unpaid balance.

Intelligent optimism is admirable, but fact is better than fiction; and blind bigotry paraded as optimism is dangerous and condemnable. Some one has said that such a bigot is not an optimist but a "cheerful idiot." To purchase rich, well-watered land at a low price and become wealthy by merely waiting till the land increases in value tenfold, while making a living by taking fertility from the soil, has been easy and common in the great agricultural states during the last half-century. But, paradoxical as it may seem, land values have increased while fertility and productiveness have decreased and, with shorter days for higher priced and less efficient farm labor, with more middlemen absorbing the profits between the producer and the consumer, it is now difficult indeed to buy land with borrowed money and pay for it from subsequent farm profits. If continued soil depletion is practiced, ultimate failure is the only future for such investments.

That vast areas of land once cultivated with profit in the original thirteen states now lie agriculturally abandoned is common knowledge; and that the farm lands of the great Corn Belt and Wheat Belt of the North-Central states are even now undergoing the most rapid soil depletion ever witnessed is known to all who possess the facts. Unless this tendency is checked these lands will go the way of the abandoned farms.

Some Broad Facts

The United States Bureau of the Census reports that the total production of our five great grain crops—corn, wheat, oats, barley and rye—amounted to 4,414,000,000 bushels in 1899, and to 4,445,000,000 bushels in 1909, an increase of less than one per cent. Furthermore, if we assume the average production reported by the United States Department of Agriculture for the three-year periods 1898 to 1900 and 1908 to 1910 as the normal for 1899, and 1909, respectively, and compare these averages with the production actually reported by that department for 1899 and 1909, we find that as an average of all these crops 1909 was a slightly more favorable season than 1899, which indicates that with strictly comparable seasons the increase from 1899 to 1909 was less than 1/2 per cent in the production of these five great grain crops of the United States.

On the other hand, the Bureau of Census reports that during the same decade the acreage of farm land in the United States increased by 4.8 per cent, and that the acreage of improved farm land-that is, farmed land-increased by 154 per cent. Thus the census data plainly show reduced yield per acre. In addition we have actual records which show that during the decade our wheat exports decreased from 210,000,000 to 108,000,000 bushels, and that our corn exports decreased from 196,000,000 to 49,000,000 bushels, in order to help feed the increase of 21 per cent in our population. And yet the people complained of the high cost of plain living and many have been forced to adopt lower standards for the table. Meanwhile the value of the farm land in the United States increased by 118 per cent during the ten years—from $13,000,000,000 to $28,500,000,000—as reported by the Bureau of Census.

The Value of Land

The great primary reason why land values have increased so markedly during the last thirty years is that America has no more free land of good quality in humid sections. Civilized man is characterized by hunger for the ownership of land. Our population continues to increase by more than 20 per cent each decade, but all future possible additions to the farm lands of the United States amount to only 9 per cent of the present acreage, and most of this small addition requires expensive irrigation or drainage.

If it cost $4 an acre to raise corn, 5 cents a bushel to harvest and market the crop, 9 cents a bushel to maintain the fertility of the soil, and 1/2 per cent on the value of the land for taxes, then, if money is worth 5 per cent, land that produces 20 bushels of 40-cent corn is worth $21.81 an acre. On the same basis, what would land be worth that produces 40 bushels of corn and equivalent values of other crops? At first thought one might say, $43.62; but this answer would be far from the correct one, which is $116.36.

And, if we again double the yield, making it 80 bushels an acre, the value of the land becomes not $87.24, and not $232.72; but easy computation will show that the gross receipts from an 80-bushel crop will pay $7.20 an acre for soil enrichment, $4 for raising the crop, $4 for harvesting and marketing, $1.53 for taxes and 5 per cent interest on a valuation of $305.45 an acre.

The average yield of corn in the United States is only 25 bushels an acre, and the average net returns even from the farms of the Corn Belt will not pay 4 per cent interest on their present market value. But the intelligent investment of $2 an acre annually in positive soil enrichment will increase the crop yield by two bushels of corn each year—or by equivalent amounts of other crops grown in the rotation—and will maintain this increase for at least a dozen years on the average land now under cultivation in the United States; and no other safe investment can be named that will pay so great returns. Of course, the cost is $1 a bushel for the first year's increase, and even the second year the 4 bushels of corn cost $2; but what is the cost per bushel of the increase the tenth year? It is 10 cents; and the twelfth year the 24 bushels of increase cost only 8-1/3 cents a bushel, with a return of nearly 500 per cent on the annual investment in soil improvement.

And this is not based on mere theoretical considerations. The average Corn-Belt land is producing only 40 bushels of corn to the acre; while a six-year average yield of 90 bushels has been produced on the common Corn-Belt land with proper and profitable soil treatment. Thus is it too much for any farmer to adopt a definite system based upon established practical scientific information which makes it possible for his yield to increase from 40 bushels to an average of 64 bushels an acre? But let him make sure that the system he adopts is cumulative and truly permanent, and not merely stimulating and temporary.

What Phosphorus Did on One Farm

On his 500-acre farm near Gilman, in the heart of the Illinois Corn Belt, Mr. Frank I. Mann has produced a 70-bushel average yield of corn for a five-year period, and with 200 acres of land in corn annually. It cost him only $1 an acre a year in fine-ground natural rock phosphate to produce increased yields of 16 bushels more corn, 23 bushels more oats and 1 ton more clover than the average yields secured without adding phosphorus.

But this progressive, practical farmer is only putting into profitable practice the results of the long-continued careful investigations with raw phosphate conducted by such public-service institutions as the agricultural experiment stations of Pennsylvania, Maryland, Rhode Island, Massachusetts, Ohio and Illinois. He knows also that on four different fields of typical Corn-Belt land in McLean county, Illinois, the total crop values per acre for a period of ten years were $148.75 $151-30, $149.43 and $149.96, respectively, and that on four other adjoining or intervening fields, which differed only by two liberal additions of phosphorus during the ten years, the respective crop values for the same time were $229,37, $221.30, $229.20 and $225.57.

Of course, Mr. Mann does not buy nitrogen, but he takes it from the inexhaustible supply in the air by means of clover and alfalfa or other legumes. He does not buy potassium because he knows how to liberate it from the inexhaustible supply contained in the soil, and because he knows that in the Illinois investigation just cited the crop values from four different fields not receiving potassium were $148.75, $151.30, $229.37 and $221.30; while four other adjoining fields, which differed only by liberal applications of potassium, produced during the same ten years $149.43, $149.96, $229.20 and $225.57, respectively.

Thus, as a general average, phosphorus increased the crop values by $76.50 an acre, which amounts to more than 300 per cent on the investment, and at the end of the ten years the soil on the best treated and highest yielding land was 10 per cent richer in phosphorus than at the beginning; while the crops from the unfertilized land removed an amount of phosphorus equal to nearly one-tenth of the total supply in the plowed soil. But a similar general average shows that potassium produced increased crop values worth only 86 cents, or 3 per cent of its cost.

What other results should be expected from land containing in the plowed soil of an acre less than 1200 pounds of phosphorus and more than 36,000 pounds of potassium?

"Working" the Land

If there is one agricultural fact that needs to be impressed upon the American people it is that the farmers of this country have been living, not upon the interest from their investments, but upon their principal; and whatever measure of apparent prosperity they have had has been taken from their capital stock. The boastful statement sometimes made, that the American landowner has become a scientific farmer, is as erroneous as it is optimistic. Such statements are based upon a few selected examples or rare illustrations, and not upon any adequate knowledge of general farm practice. Even to this date almost every effort put forth by the mass of American farmers has resulted in decreasing the fertility of the soil.

The productive power of normal land in humid climates depends almost wholly upon the power of the soil to feed the crop; but the American farmer does everything except to restore to the soil the plant food required to maintain permanently its crop-producing power. These ought be to have done, but not to leave the other undone. Thus, tile drainage adds nothing to the soil out of which crops are made, but only permits the removal of more fertility in the larger crops produced on the well-drained land. More thorough tillage with our improved implements of cultivation is merely "working the land for all that's in it." The use of better seed produces larger crops, but only at the expense of the soil. Even the farm manure is so limited and is spread so thinly with manure-spreaders made for the purpose that it adds but little to the soil in comparison with the crops removed and sold in grain and hay as well as in meat and milk. Clover, as commonly produced and harvested, adds little or no nitrogen to the soil.

The ordinary high-priced, manufactured, acidulated, so-called "complete" commercial fertilizers, in the small amounts that farmers can afford to use, and do use quite generally in the older states, serve in part as soil-stimulants and commonly leave the land poorer year by year; and if the farmers of the great Corn and Wheat Belts are ever to adopt systems of permanent agriculture, it must be done in the near future, or they too will awake to find their lands impoverished beyond self-redemption.

Even in the state of Massachusetts, where a most active campaign has been waged for forty years by the mixed commercial fertilizer interests, urging and persuading many farmers to use their high-priced artificial soil stimulants, very large areas of land are being agriculturally abandoned. Thus the following statement appears in the report of the United States Bureau of Census in regard to the farm land of Massachusetts:

"The area of improved land decreased without interruption until in 1910 it was only about one-half what it was in 1880."

It should not be forgotten, however, that market gardeners often sell from $100 to $300 worth of produce from an acre and they can well afford to use large amounts of soluble commercial plant food (acid phosphate, nitrates, etc.) as well as animal manures from the cities.

Is the Soil Inexhaustible?

It is not the fault of the farmer alone that soil-robbing and land ruin have followed his work in America. Neither the average farmer of today nor any of his ancestors received any agricultural instruction in the schools; and the greedy fertilizer agent has persuaded him to buy his patent soil medicine and has taken $100 of the farmer's money and given him in return only $10 worth of what he really needs to buy; and even the Bureau of Soils of the Federal Government has for several years promulgated the erroneous and condemnable theory expressed in the following quotations:

"From the modern conception of the nature and purpose of the soil it is evident that it cannot wear out; that, so far as the mineral food is concerned, it will continue automatically to supply adequate quantities of the mineral plant foods for crops." (United States Bureau of Soils, Bulletin No. 55, p. 79.)

"There is another way in which the fertility of the soil can be maintained: namely, by arranging a system of rotation and growing each year a crop that is not injured by the excreta of the preceding crop: then when the time comes round for the first crop to be planted again, the soil has had ample time to dispose of the sewage resulting from the growth of the plant two or three years before." (United States Farmers' Bulletin No. 257, p. 21.)

"The soil is the one indestructible, immutable asset that the nation possesses. It is the one resource that cannot be exhausted; that cannot be used up." (United States Bureau of Soils, Bulletin No. 55, p. 66.)

And these are only samples of the false teaching spread abroad by this bureau of theorists, even though the congressmen of the United States can not enter the capitol of the nation from any direction without passing depleted and agriculturally abandoned lands. Is it not in order to ask the Congress or the president of the United States how long the American farmer is to be burdened with these pernicious, disproved and condemnable doctrines poured forth and spread abroad by the Federal Bureau of Soils?

It is true that these erroneous teachings have been opposed or ridiculed in Europe; they have been denounced by the Association of Official Agricultural Chemists of the United States, and rejected by every land-grant college and agricultural experiment station that has been heard from, including those in forty-seven states; and yet this doctrine, emanating from what should be the position of highest authority, is the most potent of all existing influences to prevent the proper care of our soils.

The Values in Land

It was Baron von Liebig who taught, both in Germany and in England, that—"it is not the land itself that constitutes the farmer's wealth, but it is in the constituents of the soil, which serve for the nutrition of plants, that this wealth truly consists." And it is in the application of this teaching, completely verified by sixty years of investigation and demonstration by Lawes and Gilbert at Rothamsted, that England has been able to raise her 10-year average yield of wheat to 37-1/2 bushels an acre, while the average for the United States stands at 14 bushels.

In Illinois, where the agricultural college and experiment station, the state farmers' institute and the agricultural press have been working in perfect co-operation in teaching and demonstrating the need and value of soil enrichment as well as of seed selection and proper tillage, the 10-year average yield of wheat is already 3 bushels higher and the 10-year average yield of corn is 7-1/2 bushels higher than the averages for the 25-year period ending with 1890, before the definite information from Illinois investigations began to be widely disseminated; and yet it must be confessed that on the average Illinois is producing only 16 bushels of wheat and 36 bushels of corn to the acre, which is less than half a crop, measured by the possibilities of our soil and climate.

But what shall we say of Georgia, both an older and a larger state, and with far better climatic conditions for corn, yet with a 10-year average yield of less than 12 bushels of corn to the acre, notwithstanding the yearly expenditure of $20,000,000 for more than 2000 different brands of commercial fertilizers that have been bought by Georgia farmers? The facts are that while some profit can be secured from the use of high-priced mixed commercial fertilizers for cotton with lint at 10 cents a pound, they scarcely pay their cost when used for corn, even at Georgia prices.

Working Mind and Muscle

But Georgia spends money enough for fertilizers to double the average crop yields of the entire state within a decade if wisely invested in positive soil enrichment in rational permanent systems of agriculture.

Why should not the farmers of Georgia and other Southern states be brought to understand and to apply the results of those most valuable investigations conducted by the Louisiana Experiment Station on typical worn upland soil of the South, which show that the use of organic manures produced upon the farm-farm manure, legume cover-crops and cottonseed meal—re-enforced by liberal additions of phosphorus, increased the crop yields from 466 to 1514 pounds per acre of seed cotton, from 9.4 to 31.4 bushels of corn, and from 16.4 to 41.8 bushels of oats, as the averages for nineteen years?

This experiment occupied 6 acres of land, but when the results are applied to a 60-acre farm it is found that the gross returns from the untreated land would amount to $595.76, while the net returns from the soil treatment amount to $956.08 annually, both the value of produce and the cost of fertilizer being computed at the prices that were used by the Louisiana Experiment Station.

Thus the combined *gross* earning power of both land and labor is less than $600 a year; while the brain work applied to the improvement of the soil on the same farm brings a net return of more than $950. Once in three years 50 pounds an acre of kainit was also applied. This would contain only 5 pounds of potassium, or less than would be required for one 7-bushel crop of corn.

These are the oldest experiments in the United States in which organic manures have been re-enforced with phosphorus, and the only addition suggested for the profitable improvement of this system is ground limestone on acid soils. These results only emphasize the fact that the average farm yields small returns upon the capital and labor invested, but the statement may well be repeated that the intelligent improvement of his soil, in systems of permanent agriculture, is the most profitable business in which the farmer and land owner can engage.

www.ingramcontent.com/pod-product-compliance
Lightning Source LLC
LaVergne TN
LVHW041805190726
843493LV00008B/2799